Discovering Dinosaurs

by Janine Scott

Content and Reading Adviser: Joan Stewart
Educational Consultant/Literacy Specialist
New York Public Schools

Spyglass
BOOKS

COMPASS POINT BOOKS

Minneapolis, Minnesota

Compass Point Books
151 Good Counsel Drive
P.O. Box 669
Mankato, MN 56002-0669

Visit Compass Point Books on the Internet at *www.compasspointbooks.com*
or e-mail your request to *custserv@compasspointbooks.com*

Photographs ©: Visuals Unlimited/Jeff Greenberg, cover; Visuals Unlimited/Arthur Gurmankin and Mary Morina, 4; Visuals Unlimited/Ken Lucas, 5; Visuals Unlimited/Martin Miller, 6; Visuals Unlimited/Jana R. Jirak, 7; Visuals Unlimited/Ken Lucas, 8, 9, 10, 11, 13; Visuals Unlimited/A. J. Copley, 14; Visuals Unlimited/John D. Cunningham, 15; Visuals Unlimited/Kjell B. Sandved, 16; Visuals Unlimited/Bud Nielsen, 17; Visuals Unlimited/Jeff Greenberg, 18; Visuals Unlimited/David G. Campbell, 19; Two Coyote Studios/Mary Walker Foley, 20, 21.

Project Manager: Rebecca Weber McEwen
Editor: Alison Auch
Photo Researcher: Jennifer Waters
Photo Selectors: Rebecca Weber McEwen and Jennifer Waters
Designer: Mary Walker Foley

Library of Congress Cataloging-in-Publication Data
Scott, Janine.
  Discovering dinosaurs / by Janine Scott.
    p. cm. -- (Spyglass books)
Includes bibliographical references and index.
  ISBN-13: 978-0-7565-0231-7 (hardcover)
  ISBN-10: 0-7565-0231-4 (hardcover)
  ISBN-13: 978-0-7565-1041-1 (paperback)
  ISBN-10: 0-7565-1041-4 (paperback)
  1. Dinosaurs--Juvenile literature. [1. Dinosaurs.] I. Title. II.
Series.
  QE861.5 .S4 2002
  567.9--dc21
                          2001007333

Printed in the United States of America in North Mankato, Minnesota.
022019
005699R

Contents

Dinosaur Names 4

Dinosaur Discovery 6

Dinosaur Diets.................8

Dinosaur Defenses 10

Dinosaur Families...................12

Dinosaur Eggs 14

Dinosaur Detectives 16

Displaying Dinosaurs 18

Make Dinosaur Tracks 20

Glossary 22

Learn More 23

Index 24

Dinosaur Names

Dinosaurs were large *reptiles* that lived millions of years ago. Their names can be hard to say. Most dinosaur names describe the dinosaur.

Did You Know?
Triceratops means "three-horned face."

Triceratops

Dinosaur Discovery

Dinosaurs have been gone for 65 million years, but people have not known about them for very long. In 1841, scientists decided that the large bones belonged to animals they called *Dinosauria*. This means "terrible lizard."

Bone *fossil*

Did You Know?
Before 1841, when people found bones of dinosaurs, some thought they were bones of giant humans!

Fossils

Dinosaur Diets

Some dinosaurs ate plants.
Some ate plant-eating dinosaurs.

Plant-eating dinosaurs
often had long necks
for reaching leaves in trees.
Meat eaters had short necks,
strong jaws, and sharp teeth
for ripping and chewing meat.

Tyrannosaurus rex

Dinosaur Defenses

Different dinosaurs had different *defenses* against enemies.

Some had spiked tails, horns, hard plates, and claws. Some were fast runners.

Did You Know?

Some dinosaurs were so large they could not run away. Their size was their protection.

Stegosaurus

Dinosaur Families

Some dinosaurs, such as Triceratops, lived in **herds**. The herd gave them protection against meat-eating dinosaurs.

Some larger, meat-eating dinosaurs, such as the Tyrannosaurus rex, did not need the protection of the herd.

Did You Know?

Scientists think some dinosaurs took care of their young after the eggs hatched.

Baby dinosaurs and eggs

Dinosaur Eggs

Dinosaurs laid their eggs in hollows in the ground. The eggs had leathery shells and were oval-shaped.

Baby dinosaur

Did You Know?

One dinosaur nest has been found that had thirty-seven eggs!

Dinosaur eggs

Dinosaur Detectives

Scientists hunt for dinosaur remains all over the world.

Dinosaur fossils give scientists or "dinosaur *detectives*" clues about the dinosaurs' way of life.

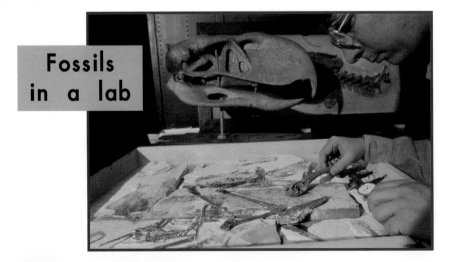

Fossils in a lab

Did You Know?

Sometimes scientists wrap fossils in special "jackets" after they find them. These protect the fossils before they are taken back to a laboratory.

Fossils in the ground

Displaying Dinosaurs

In the past, museums simply displayed dinosaur fossils. Today, some museums have dinosaur robots and videos of "dinosaur detectives" at work.

Dinosaur footprint

Did You Know?

"Sue" is the name of the world's most complete Tyrannosaurus rex skeleton. She was found in South Dakota. It took 25,000 hours to clean and put together her bones!

Tyrannosaurus rex

Make Dinosaur Tracks

You will need:
- Modeling clay

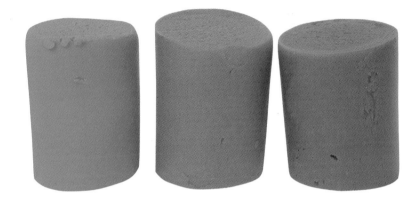

1. Mash the clay as flat as a plate.

2. Using your fingers, make tracks that "walk" across the clay.

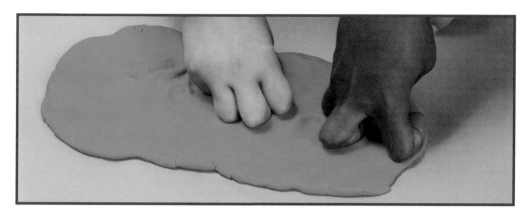

3. Let the clay dry. Now you have your own dinosaur fossil!

Glossary

defense—something an animal uses to keep out of danger

detective—a person who collects clues to figure something out

fossil—the remains of ancient plants and animals

herd—a group of the same kind of animals that live together

reptile—an animal that breathes air and is usually covered with scales or plates

Learn More

Books

Aliki. *Dinosaurs Are Different*. New York: Crowell, 1985.

Gibbons, Gail. *Dinosaurs*. New York: Holiday House, 1987.

Parish, Peggy. *Dinosaur Time*. Illustrated by Arnold Lobel. New York: Harper & Row, 1974.

Web Sites

For more information on dinosaurs, use FactHound.

1. Go to *www.facthound.com*
2. Type in a search word related to this book or this book ID: 0756502314
3. Click on the *Fetch It* button.

FactHound will fetch the best Web sites for you!

Index

eggs, 13, 14, 15

fossils, 6, 7, 16, 17, 18

herds, 12

scientists, 6, 13, 16, 17

Stegosaurus, 10, 11

Triceratops, 4, 5, 12

Tyrannosaurus rex, 8, 9, 12, 19

GR: H
Word Count: 211

From Janine Scott

I live in New Zealand, and have two daughters. They love to read fact books that are full of fun facts and features. I hope you do, too!